EMMANUEL JOSEPH

From Nebulas to Novels, The Psychological and Literary Exploration of the Universe

Contents

1

Chapter 1: The Genesis of Cosmic Curiosity

Since time immemorial, humanity has gazed at the stars with a sense of wonder and curiosity. The mysteries of the cosmos have sparked imaginations and driven us to explore beyond our earthly confines. Ancient civilizations, from the Babylonians to the Greeks, saw the heavens as the realm of gods and mythical creatures. These early interpretations, though primitive by modern standards, laid the groundwork for a deeper understanding of the universe and our place within it.

The Renaissance era marked a significant shift in our cosmic perspective. With the advent of telescopes, astronomers like Galileo Galilei and Johannes Kepler unlocked new vistas of knowledge. Their observations challenged the geocentric model of the universe, placing the sun, not Earth, at the center. This heliocentric revelation profoundly impacted both science and society, fostering a sense of interconnectedness and humility in the face of the vast, seemingly infinite universe.

As our comprehension of the cosmos expanded, so too did our literary imaginations. Writers began to incorporate cosmic themes into their works, blending scientific curiosity with creative storytelling. The 19th century witnessed the birth of science fiction, a genre that dared to envision the future and explore the unknown. Pioneering authors like Jules Verne and H.G.

Wells crafted tales of interstellar travel, alien civilizations, and technological marvels, captivating readers and inspiring future generations of scientists and dreamers.

In the contemporary era, our fascination with the universe has only intensified. The Hubble Space Telescope, launched in 1990, provided unprecedented glimpses into the cosmos, revealing the beauty and complexity of galaxies, nebulae, and distant stars. These breathtaking images have not only expanded our scientific knowledge but also inspired a new wave of literary exploration. Modern authors continue to weave cosmic themes into their narratives, reflecting our enduring quest for understanding and connection with the universe.

2

Chapter 2: The Human Psyche and the Infinite Expanse

The vastness of the universe has a profound impact on the human psyche. When we contemplate the enormity of space, we are often confronted with feelings of awe, insignificance, and existential wonder. This psychological response, known as the "overview effect," is experienced by astronauts who view Earth from space. From this vantage point, our planet appears as a fragile, interconnected whole, prompting a shift in perspective that transcends national borders and personal concerns.

Exploring the cosmos can also evoke a sense of the sublime, a term coined by philosopher Edmund Burke to describe the awe-inspiring beauty and terror of nature. The infinite expanse of space, with its countless stars and galaxies, elicits a powerful emotional response that challenges our understanding of reality and our place within it. This sense of the sublime has long been a source of inspiration for poets, artists, and writers, who seek to capture the ineffable qualities of the cosmos in their work.

The psychological impact of the universe extends beyond feelings of awe and wonder. The search for extraterrestrial life, for example, raises profound questions about our uniqueness and the potential for other intelligent beings in the cosmos. The possibility of contact with extraterrestrial civilizations has been a recurring theme in literature, from H.G. Wells' "The War of the

Worlds" to Carl Sagan's "Contact." These narratives explore the implications of such encounters, challenging our assumptions about humanity and our place in the universe.

Our relationship with the cosmos also influences our sense of time and mortality. The vast timescales of astronomical events, from the formation of stars to the expansion of galaxies, dwarf the brevity of human life. This awareness of cosmic time can evoke feelings of humility and introspection, prompting us to reflect on the fleeting nature of existence and the enduring legacy of our actions. Literature serves as a vehicle for exploring these themes, offering readers a means of grappling with the existential questions posed by the universe.

3

Chapter 3: Celestial Imagery in Ancient Mythology

Ancient civilizations looked to the stars and saw the divine. Celestial bodies were often imbued with spiritual significance, serving as symbols of gods, heroes, and mythological events. In Greek mythology, the constellations were stories written in the night sky, mapping the adventures of characters like Orion, Perseus, and Andromeda. These narratives provided a framework for understanding natural phenomena and human experiences, linking the heavens to earthly life.

In ancient Egypt, the stars were associated with the afterlife and the eternal journey of the soul. The pharaohs were believed to become stars upon their death, joining the gods in the celestial realm. The Pyramids of Giza, aligned with the stars of Orion's Belt, served as monumental gateways to the heavens, reflecting the Egyptians' belief in the interconnectedness of the cosmos and the afterlife.

The indigenous cultures of the Americas also wove celestial themes into their mythology. The Mayans, renowned for their astronomical knowledge, developed a complex calendar system based on the movements of the sun, moon, and planets. Their creation myth, the Popol Vuh, tells of a divine struggle between the gods and the forces of chaos, culminating in the birth of the sun and the moon. These stories were not only a means of explaining the cosmos but also a way of reinforcing social and cultural values.

In ancient India, the cosmos was seen as a manifestation of the divine order. The Rigveda, one of the oldest sacred texts, describes the universe as a cosmic being, Purusha, whose sacrifice led to the creation of the world. Hindu mythology is replete with celestial imagery, from the churning of the cosmic ocean to the dance of Shiva, the cosmic destroyer and creator. These narratives reflect the cyclical nature of the universe and the eternal interplay of creation and destruction.

4

Chapter 4: The Birth of Modern Astronomy and Its Literary Reflections

The 16th and 17th centuries marked a revolution in our understanding of the cosmos. The work of astronomers like Nicolaus Copernicus, Galileo Galilei, and Johannes Kepler challenged the long-held geocentric model, placing the sun at the center of the solar system. This heliocentric view, though initially controversial, laid the foundation for modern astronomy and fundamentally altered our perception of the universe.

Galileo's observations of the moons of Jupiter and the phases of Venus provided concrete evidence for the heliocentric model, while Kepler's laws of planetary motion described the precise movements of celestial bodies. These discoveries were not only scientific milestones but also sources of inspiration for contemporary writers. John Milton's epic poem "Paradise Lost" reflects the shifting cosmic perspective of the time, incorporating astronomical imagery and references to the new scientific discoveries.

The Scientific Revolution continued to influence literature in the centuries that followed. The works of the Romantic poets, such as William Wordsworth and Samuel Taylor Coleridge, were shaped by the emerging understanding of the cosmos and the natural world. These poets sought to capture the emotional and philosophical implications of scientific discoveries, exploring themes of nature, the sublime, and the interconnectedness of all things.

The 19th century saw the rise of science fiction, a genre that allowed writers to speculate about the future and explore the unknown. Jules Verne, often considered the father of science fiction, penned imaginative tales of space travel and extraterrestrial adventures, blending scientific knowledge with creative storytelling. His works, including "From the Earth to the Moon" and "Twenty Thousand Leagues Under the Sea," captivated readers and inspired a sense of wonder about the possibilities of the universe.

5

Chapter 5: The Cosmic Sublime in Romantic Literature

The Romantic period, spanning the late 18th and early 19th centuries, was characterized by a profound appreciation for the beauty and majesty of the natural world. This era saw a resurgence of interest in the cosmos, as poets and writers sought to capture the sublime qualities of the universe in their works. The cosmic sublime, a term used to describe the awe-inspiring vastness and mystery of the cosmos, became a central theme in Romantic literature.

William Wordsworth, one of the foremost Romantic poets, often incorporated celestial imagery into his poetry. In his "Ode: Intimations of Immortality," Wordsworth reflects on the transcendent beauty of the natural world, drawing connections between the heavens and the human soul. His work emphasizes the interconnectedness of all things, suggesting that the universe is a source of spiritual inspiration and existential contemplation.

Samuel Taylor Coleridge, another key figure of the Romantic period, explored the cosmic sublime in his poem "The Rime of the Ancient Mariner." The mariner's journey across the vast and treacherous ocean serves as a metaphor for the human quest for understanding and connection with the universe. Coleridge's vivid descriptions of the natural world, combined with his exploration of the supernatural, evoke a sense of wonder and awe that is characteristic of the cosmic sublime.

The Romantic poets were not alone in their fascination with the cosmos. Mary Shelley, the author of "Frankenstein," incorporated scientific and philosophical themes into her novel, reflecting the era's fascination with the unknown. The creature's creation, influenced by contemporary scientific advancements, raises questions about humanity's quest for knowledge and the ethical implications of scientific exploration. Shelley's work serves as a cautionary tale, highlighting the potential consequences of humanity's pursuit of cosmic understanding.

The influence of the cosmic sublime extended beyond poetry and fiction. The visual arts also embraced the themes of vastness and mystery, with painters like J.M.W. Turner and Caspar David Friedrich creating landscapes that captured the grandeur and majesty of the natural world. These artists sought to convey the emotional and philosophical impact of the cosmos, reflecting the Romantic era's deep appreciation for the beauty and mystery of the universe.

6

Chapter 6: Science Fiction and the Exploration of Otherworldly Realms

The genre of science fiction has long been a vehicle for exploring the mysteries of the universe and the possibilities of otherworldly realms. From the early works of pioneers like H.G. Wells and Jules Verne to the modern masterpieces of Arthur C. Clarke and Isaac Asimov, science fiction has continually pushed the boundaries of human imagination. These authors dared to envision futures where humanity transcends its terrestrial limits, venturing into the stars and encountering alien civilizations.

H.G. Wells' "The War of the Worlds" is a prime example of how science fiction can reflect societal anxieties and aspirations. The novel's depiction of a Martian invasion serves as an allegory for the destructive potential of advanced technology and the vulnerabilities of human civilization. Wells' narrative challenges readers to consider the ethical and philosophical implications of our scientific advancements and our place in the cosmos.

Jules Verne's "From the Earth to the Moon" is another seminal work that captures the spirit of scientific curiosity and exploration. The novel's detailed description of a lunar expedition, though fictional, is grounded in the scientific knowledge of the time. Verne's work not only entertained readers but also inspired future generations of scientists and engineers to pursue the dream of space travel.

In the 20th century, science fiction continued to evolve, reflecting the rapid advancements in technology and our growing understanding of the universe. Arthur C. Clarke's "2001: A Space Odyssey" explores the themes of artificial intelligence, extraterrestrial life, and the evolution of human consciousness. The novel's enigmatic monoliths and the iconic HAL 9000 computer have become symbols of the genre's capacity to provoke profound questions about the future of humanity and our relationship with technology.

Isaac Asimov's "Foundation" series delves into the complexities of galactic empires, psychohistory, and the cyclical nature of civilizations. Asimov's intricate world-building and his exploration of the interplay between science, politics, and human behavior have left an indelible mark on the genre. Through these imaginative narratives, science fiction serves as a lens through which we can examine the possibilities and challenges of our own future.

7

Chapter 7: The Role of Astronomy in Modern Literature

The advances in modern astronomy have had a profound impact on contemporary literature, inspiring writers to explore the mysteries of the cosmos and our place within it. The discovery of exoplanets, the study of black holes, and the search for dark matter have provided fertile ground for literary exploration, blending scientific knowledge with imaginative storytelling.

The discovery of exoplanets has sparked a renewed interest in the search for extraterrestrial life. Authors like Kim Stanley Robinson and Liu Cixin have explored the implications of interstellar colonization and contact with alien civilizations. Robinson's "Aurora" and Liu's "The Three-Body Problem" delve into the challenges and ethical dilemmas of space exploration, questioning the sustainability of human expansion and the potential consequences of encountering advanced extraterrestrial species.

Black holes, with their mysterious and mind-bending properties, have also captured the literary imagination. Stephen Baxter's "The Light of Other Days" and Alastair Reynolds' "Revelation Space" series explore the concept of wormholes and the potential for faster-than-light travel. These narratives push the boundaries of scientific plausibility, challenging readers to consider the fundamental nature of space and time.

The search for dark matter and dark energy, which constitute the majority of the universe's mass and energy, has also inspired literary exploration. Works like Greg Egan's "Diaspora" and Neal Stephenson's "Anathem" delve into the mysteries of the cosmos, blending cutting-edge scientific concepts with philosophical inquiry. These authors invite readers to contemplate the hidden aspects of the universe and the limits of human knowledge.

In addition to these scientific themes, modern literature continues to explore the psychological and existential implications of our cosmic discoveries. Authors like Margaret Atwood and Kazuo Ishiguro use speculative fiction to examine the human condition in the context of a vast and indifferent universe. Atwood's "Oryx and Crake" and Ishiguro's "Never Let Me Go" address themes of identity, mortality, and the ethical implications of scientific advancement, reflecting our enduring quest for meaning and understanding in the face of cosmic uncertainty.

8

Chapter 8: The Intersection of Science and Spirituality

The exploration of the cosmos has often intersected with questions of spirituality and the search for meaning. Throughout history, the contemplation of the universe has inspired profound reflections on the nature of existence and our place within it. Modern literature continues to explore this intersection, blending scientific inquiry with spiritual themes to create narratives that resonate with readers on a deep, existential level.

Carl Sagan's "Contact" is a quintessential example of this synthesis. The novel follows the story of Dr. Ellie Arroway, a scientist who discovers a signal from an extraterrestrial civilization. Sagan's narrative explores the implications of this discovery, addressing themes of faith, skepticism, and the quest for knowledge. Through the character of Ellie, Sagan examines the delicate balance between scientific rigor and spiritual wonder, highlighting the potential for reconciliation between these seemingly disparate realms.

Arthur C. Clarke's "Childhood's End" also delves into the intersection of science and spirituality. The novel envisions a future where humanity is guided by an advanced alien civilization, leading to a transformation of human consciousness. Clarke's narrative explores the themes of transcendence, evolution, and the ultimate destiny of humanity, suggesting that our journey through the cosmos is not only a scientific endeavor but also a spiritual quest for meaning and enlightenment.

The exploration of the cosmos has also inspired philosophical reflections on the nature of reality and the limits of human knowledge. Works like Philip K. Dick's "Do Androids Dream of Electric Sheep?" and Stanisław Lem's "Solaris" challenge readers to question the nature of consciousness, identity, and perception. These narratives invite readers to consider the possibility that our understanding of the universe is shaped by our subjective experiences and that the quest for knowledge is inherently intertwined with the search for self-awareness.

In contemporary literature, authors like Haruki Murakami and Ted Chiang continue to explore the intersection of science and spirituality. Murakami's "1Q84" and Chiang's "Story of Your Life" blend elements of speculative fiction with metaphysical inquiry, creating narratives that explore the mysteries of existence and the boundaries of human understanding. These works reflect our enduring fascination with the cosmos and our quest for meaning in a universe that is vast, complex, and often elusive.

9

Chapter 9: The Influence of Space Exploration on Popular Culture

The achievements of space exploration have had a profound impact on popular culture, inspiring a myriad of films, television shows, and other forms of media. From the iconic imagery of the Apollo moon landings to the futuristic visions of "Star Trek" and "Star Wars," the exploration of the cosmos has captured the imagination of audiences around the world.

The Apollo missions, which culminated in the historic moon landing of 1969, marked a watershed moment in human history. The images of astronauts walking on the lunar surface became iconic symbols of human ingenuity and determination. This achievement not only inspired a generation of scientists and engineers but also influenced popular culture, sparking a wave of space-themed films, books, and television shows.

The science fiction television series "Star Trek," created by Gene Rodden-berry, envisioned a future where humanity explores the galaxy, encountering diverse alien civilizations and grappling with complex moral and ethical dilemmas. The show's optimistic vision of the future, with its emphasis on exploration, cooperation, and the pursuit of knowledge, resonated with audiences and has remained a cultural touchstone for decades.

The "Star Wars" franchise, created by George Lucas, took a different approach, blending elements of science fiction, fantasy, and mythology to create an epic saga set in a galaxy far, far away. The franchise's rich world-building, memorable characters, and epic storytelling have captivated audiences and inspired countless works of fan fiction, merchandise, and spin-offs. "Star Wars" has become a cultural phenomenon, reflecting our fascination with the cosmos and the timeless appeal of heroic adventure.

Space exploration has also influenced the music, art, and fashion of popular culture. The iconic imagery of space travel, from astronaut suits to spacecraft, has been incorporated into various artistic expressions. Musicians like David Bowie and Elton John have drawn inspiration from the cosmos, creating songs that capture the wonder and mystery of space. The visual arts have also embraced celestial themes, with artists like Yayoi Kusama and Olafur Eliasson creating works that evoke the beauty and vastness of the universe.

10

Chapter 10: The Role of Space Agencies and Private Enterprises

The exploration of the cosmos has been driven by the efforts of space agencies and private enterprises, each contributing to our understanding of the universe and pushing the boundaries of what is possible. Organizations like NASA, the European Space Agency (ESA), and private companies like SpaceX have played pivotal roles in advancing space exploration and inspiring the public imagination.

NASA, the United States' National Aeronautics and Space Administration, has been at the forefront of space exploration since its inception in 1958. The agency's achievements, from the Apollo moon landings to the Mars rover missions, have expanded our knowledge of the cosmos and demonstrated the potential for human ingenuity and perseverance. NASA's missions have also inspired countless works of literature, film, and other media, reflecting the public's fascination with space exploration.

The European Space Agency (ESA) has also made significant contributions to our understanding of the universe. The ESA's missions, such as the Rosetta comet mission and the Gaia space observatory, have provided valuable data on the composition and dynamics of celestial bodies. The agency's collaboration with other space organizations, including NASA and the Russian space agency Roscosmos, has fostered a spirit of international cooperation and shared scientific discovery.

Private enterprises, such as SpaceX, founded by Elon Musk, have introduced a new era of space exploration. SpaceX's achievements, including the development of reusable

continue

Private enterprises, such as SpaceX, founded by Elon Musk, have introduced a new era of space exploration. SpaceX's achievements, including the development of reusable rockets and the successful launch of crewed missions to the International Space Station (ISS), have demonstrated the potential for cost-effective and sustainable space travel. The company's ambitious goals, such as the colonization of Mars, have captured the public's imagination and inspired a renewed interest in the possibilities of human space exploration.

The collaboration between space agencies and private enterprises has led to significant advancements in space technology and exploration. The development of the Artemis program, a collaboration between NASA and private companies, aims to return humans to the moon and establish a sustainable presence on the lunar surface. This program represents a new chapter in space exploration, with the potential to pave the way for future missions to Mars and beyond.

The role of space agencies and private enterprises extends beyond the realm of science and technology. Their achievements serve as symbols of human ingenuity and determination, inspiring generations of scientists, engineers, and explorers. The stories of these endeavors, both real and imagined, continue to captivate audiences and fuel our collective fascination with the cosmos.

11

Chapter 11: The Impact of Space Exploration on Society and Culture

The exploration of the cosmos has had a profound impact on society and culture, shaping our understanding of the universe and our place within it. The achievements of space exploration have inspired a sense of wonder and possibility, encouraging us to look beyond our immediate concerns and consider the broader context of our existence.

The Apollo moon landings, for example, were not only a technological triumph but also a cultural milestone. The sight of astronauts walking on the lunar surface was a powerful symbol of human achievement and the potential for exploration and discovery. This event, broadcast to millions around the world, fostered a sense of global unity and shared accomplishment, transcending political and ideological divisions.

The advancements in space technology have also had practical applications that impact our daily lives. Satellite technology, developed for space exploration, has revolutionized communication, navigation, and weather forecasting. The Global Positioning System (GPS), initially developed for military use, has become an essential tool for transportation, commerce, and personal navigation. These technological innovations, born from our quest to explore the cosmos, have had far-reaching benefits for society.

The cultural impact of space exploration extends to the arts and humanities, inspiring a myriad of creative expressions. From the iconic images of space missions to the imaginative narratives of science fiction, the cosmos has served as a rich source of inspiration for artists, writers, and filmmakers. These cultural works not only entertain and engage audiences but also provoke reflection on the possibilities and challenges of our exploration of the universe.

Space exploration has also raised important ethical and philosophical questions about our responsibility to the cosmos and to each other. The potential for planetary colonization, the search for extraterrestrial life, and the preservation of celestial bodies all prompt us to consider the ethical implications of our actions. These questions are reflected in literature, film, and other media, challenging us to think critically about our place in the universe and our responsibilities as stewards of the cosmos.

12

Chapter 12: The Future of Cosmic Exploration and Human Imagination

As we look to the future, the exploration of the cosmos promises to continue shaping our understanding of the universe and our place within it. The advancements in space technology, the discoveries of new celestial bodies, and the potential for human missions to Mars and beyond all point to an exciting and transformative era of exploration.

The potential for human missions to Mars represents one of the most ambitious goals of contemporary space exploration. Organizations like NASA and SpaceX are working towards the development of the technology and infrastructure needed to sustain human life on the Red Planet. The challenges of such a mission, from the long-duration space travel to the harsh Martian environment, are immense, but the potential rewards in terms of scientific discovery and human achievement are equally significant.

The search for extraterrestrial life remains a central focus of cosmic exploration. The discovery of microbial life on Mars, the exploration of the icy moons of Jupiter and Saturn, and the detection of potentially habitable exoplanets all fuel our quest to answer the age-old question of whether we are alone in the universe. The implications of such a discovery would be profound, challenging our understanding of biology, evolution, and the uniqueness of life on Earth.

The future of cosmic exploration also holds the promise of new technological innovations and scientific breakthroughs. The development of advanced propulsion systems, the study of gravitational waves, and the exploration of dark matter and dark energy all represent the frontiers of contemporary astrophysics. These advancements have the potential to revolutionize our understanding of the universe and our place within it.

The exploration of the cosmos will continue to inspire human imagination and creativity. The stories we tell about the universe, from scientific discoveries to speculative fiction, reflect our enduring curiosity and our quest for meaning. As we venture further into the cosmos, we carry with us the hopes, dreams, and aspirations of generations past and present, forging a connection between the stars and the stories we create.

13

Epilogue: The Endless Frontier

The journey from nebulas to novels, from the vast expanse of the cosmos to the intricate narratives of human imagination, is a testament to our enduring curiosity and creativity. The exploration of the universe, both through scientific inquiry and literary expression, reflects our deepest desires to understand the world around us and our place within it.

The cosmos serves as both a physical frontier and a metaphorical canvas, inspiring us to explore the unknown and to create new stories that capture the wonder and mystery of the universe. As we continue to push the boundaries of our knowledge and imagination, we are reminded that the journey of exploration is as important as the discoveries we make along the way.

In the end, the exploration of the cosmos is a reflection of our humanity—our quest for knowledge, our desire for connection, and our drive to create meaning in an ever-expanding universe. The stories we tell, from scientific discoveries to imaginative tales, are an integral part of this journey, weaving a tapestry of wonder and discovery that stretches from the nebulas to the novels of our collective imagination.

From Nebulas to Novels: The Psychological and Literary Exploration of the Universe

In this intriguing exploration, "From Nebulas to Novels" delves into the profound connections between our understanding of the cosmos and the creative impulses that drive human expression. Through twelve thought-provoking chapters, the book bridges the vast expanse of the universe with the intricate realms of psychology and literature.

Beginning with the historical roots of cosmic curiosity, the book traces humanity's fascination with the stars from ancient mythology to modern scientific discoveries. Each chapter weaves together the psychological impact of cosmic exploration and its influence on literary imagination, revealing how our quest to understand the universe shapes our collective consciousness and creative endeavors.

Readers will journey through the transformative effects of the "overview effect" experienced by astronauts, the sublime beauty of the cosmos as captured by Romantic poets, and the futuristic visions of science fiction authors. The book also examines the role of space exploration in popular culture, the intersection of science and spirituality, and the ethical implications of our cosmic pursuits.

"From Nebulas to Novels" is a compelling narrative that invites readers to contemplate the mysteries of the universe and the stories we create in our quest for meaning. It is a celebration of human curiosity, creativity, and the enduring connection between the stars and the tales we tell